Sisters in STEM
Charting Paths to Innovation

Abraham George Parsons

Table of Contents

1. Introduction 2
2. Inspiring Beginnings: Women Pioneers in STEM 3
   2.1. Venturing into Uncharted Realms: Women in Science 3
   2.2. The Digital Voyage: Women in Technology 4
   2.3. Structural Mapmakers: Women in Engineering 5
   2.4. Decoding the Universe: Women in Mathematics 5
3. Breaking Barriers: The Triumph Over Historically Male Dominance 7
   3.1. Setting the Stage for Change 7
   3.2. In STEM We Trust 7
   3.3. Triumph Amidst Turbulence 8
   3.4. An Unceasing March 8
4. The Glass Ceiling: An Obstinate Reality In STEM 10
   4.1. Unraveling the Concept: The Glass Ceiling 10
   4.2. Historical Context 10
   4.3. Gender Stereotypes: Fuelling the Fire 11
   4.4. Biases in the Workplace: Subtle yet Significant 11
   4.5. The Vicious Circle: Few Women in Top Positions Propagate the Trend 12
   4.6. The Economic Impact: Pay Gap and More 12
   4.7. Charting Forward: Potential Measures 12
5. The STEM Gender Gap: Dissecting Its Origins and Effects 14
   5.1. Historical Context of Gender Bias 14
   5.2. Socialization and Its Effect on Interest Formation 15
   5.3. Gender Stereotypes and Bias in Educational Institutions 15
   5.4. Impacts of the STEM Gender Gap 16
   5.5. Conclusion: A Call to Bridge the STEM Gender Gap 16
6. Role Models and Mentors: Their Pervasive Influence in STEM

    Career Paths 17
        6.1. The Role Model Effect: Inspiring Generations 17
        6.2. The Transformational Influence of Mentors 18
        6.3. The Empowerment of Peer Networks 19
        6.4. The Imperative of Representation: 'You Can't Be What You Can't See' 19
7. Indian Women in STEM: A Success Story 21
        7.1. Industry Overview and Background 21
        7.2. Pioneering Presence: Women Who Shattered The Stereotypes 22
        7.3. Present Scenario: Growth, Trends and Challenges 22
        7.4. Way Forward: Policies, Initiatives, and Future Prospects 23
8. African Queens of STEM: From Cape To Cairo 25
        8.1. The Illuminating Torchbearers 25
        8.2. Confronting the Whirlwind: Challenges and Barriers 26
        8.3. Shaping the Future: Reformative Measures and Initiatives .. 26
        8.4. Stories of Success: Blossoming Despite the Odds 27
9. Putting Latinas on the STEM Map: Triumphs and Challenges 29
        9.1. Building up the Numbers: The Current Landscape 29
        9.2. Challenging the Socio-Cultural Paradigm 30
        9.3. Unveiling the Triumphs 30
        9.4. Bridging the Gap: An uphill Task 30
        9.5. Looking Forward: A Pledge for Inclusion 31
10. Future Endeavors: Emerging Fields And Opportunities For Women In STEM 33
        10.1. The Advent of Artificial Intelligence and Machine Learning 33
        10.2. The Rise of Data Science 34
        10.3. The Revolutionary Biotechnology and Genomics 34
        10.4. The Expanding Universe of Cybersecurity 35

10.5. The Ever-evolving Domain of Quantum Computing 35
11. Creating Inclusive Spaces: Recommendations for Encouraging Women in STEM 37
    11.1. Encouraging Early Interest in STEM 37
    11.2. Addressing Stereotypes and Biases 38
    11.3. Role of Mentorship 38
    11.4. Government and Scholarly Interventions 38
    11.5. Inclusive Hiring Practices and Job Security 39
    11.6. Bridging the Confidence Gap 39

We need to encourage and support girls and women achieve their full potential as scientific researchers and innovators.

— António Guterres

Chapter 1. Introduction

In our Special Report, "Sisters in STEM: Charting Paths to Innovation," we delve into the impressive narratives of women making waves in the Science, Technology, Engineering, and Mathematics fields. The journey is a medley of triumphs, trials, and pure resilience, often facilitated by personal determination and a supportive community. This report illuminates the strides women have made, along with a hard-nosed examination of the barriers still hampering further progress. It's not just a series of stories, but a clarion call—to educators, policymakers, and future STEM stars—to foster an environment where every spark of potential can light up the path of innovation. Discover, celebrate, and get inspired with "Sisters in STEM," a report guaranteed to stir your curiosity and enthrall your intellectual appetite, thus making it an absolute must-have!

Chapter 2. Inspiring Beginnings: Women Pioneers in STEM

Each letter in the abbreviation 'STEM' stands for a field that is profound, expansive, and responsible in its own way for driving the wheels of human progression. Science: the quest for understanding, illuminating the mysteries of the universe. Technology: the application of scientific discoveries to improve and simplify human life. Engineering: the act of intertwining creativity and logic to erect infrastructures that elevate our civilizations. Mathematics: the language spoken by the universe, itself an intricate tapestry of numbers, patterns, and calculations. Now, imagine the extraordinary resolve required of women in a patriarchal society to etch their names into the annals of these fields. This chapter aims to commemorate these remarkable beginnings, offering a heartfelt tribute to those audacious women pioneers who were audacious and fearless in their pursuits, making indelible marks in the domains of Science, Technology, Engineering, and Mathematics.

2.1. Venturing into Uncharted Realms: Women in Science

Women's contributions to science span centuries, and these pioneers, often unsung, have defied societal, institutional, and personal challenges to blaze trails in various scientific domains. One of the most iconic figures in this regard is surely Marie Curie. The Polish-born scientist's monumental discoveries in the realm of radioactivity—a term she herself coined—earned her not one, but two, Nobel Prizes in different sciences, a feat unmatched to this day. She paved the way for future women in science through her courageous determination and inextinguishable thirst for knowledge.

Fast forward through time, to the modern era, we have Barbara McClintock, an illustrious American geneticist whose research on genetic recombination in maize plants led to ground-breaking discoveries about gene regulation and transposons. Her efforts garnered her a Nobel Prize in Physiology or Medicine in 1983. These women, among numerous others, not only enhanced our understanding of the world but inspired generations of women scientists in the making.

2.2. The Digital Voyage: Women in Technology

Historically, women have played roles as integral as men's in advancing our technological accomplishments. Starting from Ada Lovelace, known as the world's first computer programmer, who envisioned the future potential of Charles Babbage's Analytical Engine in the 19th century. Her notes on the device—deemed an anachronistic fascination in her time—ultimately became the foundation for modern computing, a demonstration of her exceptional foresight.

Similarly, the name Grace Hopper resonates powerfully within computer science circles. An American computer scientist and United States Navy rear admiral who is credited with developing COBOL—one of the earliest high-level programming languages—Hopper's contributions to software development concepts continue to influence technology today.

2.3. Structural Mapmakers: Women in Engineering

The world of engineering, a bastion largely held by men, saw its fortifications breached by the relentless pursuits of women engineers. Elsie Eaves, the first female associate member of the American Society of Civil Engineers, is one iconic figure. Her role in promoting women's participation in civil engineering laid the foundation for future generations to break boundaries.

Stephanie Kwolek, a chemist by training, also warrants deserving mention in this context. Her ingenuity led to the invention of Kevlar, the high-strength material five times stronger than steel, extensively used in bulletproof vests, saving countless lives to date.

2.4. Decoding the Universe: Women in Mathematics

Mathematics is considered the language of science, and women, despite facing manifold detractions, have authored some of its compelling narratives. Take, for instance, the story of Emmy Noether, an influential German mathematician known for her groundbreaking contributions to abstract algebra and theoretical physics. Considered by Albert Einstein and many others as the most important woman in the history of mathematics, her theorems are foundational to 20th-century physics.

Further, Maryam Mirzakhani, an Iranian mathematician and a professor of mathematics at Stanford University, became the first and, to date, only female winner of the prestigious Fields Medal, the highest award in mathematics. Her complex and path-breaking work on the dynamics and geometry of mathematical objects called Riemann surfaces made seismic waves in the field.

True pioneers, these 'women of firsts' not only delved deep into their respective fields but they also surfaced, holding torches high, radiating the path for their successors. These are the inspiring beginnings—the hopes, triumphs, and the unyielding grit underscoring their journeys. Their examples continue to remind us that while women have made great strides in STEM, there remain many milestones to conquer. Undeniably, though, their indomitable spirit signals a beacon of hope for more balanced, inclusive, and diverse STEM fields in decades to come. A newly dawned era where potential, not gender, will be the compelling force behind innovation—a future all of us must aspire and strive for.

Chapter 3. Breaking Barriers: The Triumph Over Historically Male Dominance

Historical scenarios have largely characterized STEM as an area of male dominance. Women were sidelined, their potential untapped, with patriarchy underlining societal norms. However, the last century witnessed a surge in women emerging undeterred from the shadows, breaking these historical barriers with unflinching courage and determination.

3.1. Setting the Stage for Change

The onset of the twentieth century ignited the spark of change. Muted voices were voicing dissent, challenging existing societal norms. Physicist and chemist Marie Curie was a beacon of hope during this era. Despite restrictions and biases, she achieved two Nobel Prizes in different sciences – a feat that challenged the existing norm about ability and gender. Her triumph didn't just pave the way for women but shook the world of science, gently dismantling the gender paradigm.

In parallel ran the story of computer science pioneer Grace Hopper. She is renowned for developing one of the first compilers and for her instrumental role in developing COBOL, an English-like programming language. Despite the industry being predominantly male, Hopper's contributions were indispensable, leading to her nickname, "the mother of COBOL."

3.2. In STEM We Trust

The rise in women's contributions was steady if slow. In 1963, Maria

Goeppert Mayer won the Nobel Prize for Physics, leaving an indelible imprint. Her work on the nuclear shell structure of atoms has been fundamental in scientific understandings, resonating with the world that women's intellect knows no bounds.

Then came the Space Age, and with it, a remarkable woman named Katherine Johnson. Surpassing racial and gender boundaries, she was instrumental in NASA's lunar project, her mathematical calculations enabling man's first landing on the moon. For years, her story remained untold, until recently, when it received well-deserved recognition in "Hidden Figures."

Further progress was clocked in the 1980s, with Barbara McClintock's work on "jumping genes" earning her a Nobel Prize. Her findings dismissed the static nature of genes assumption, opening fresh avenues of genomic studies and rewriting textbooks.

3.3. Triumph Amidst Turbulence

Each triumph was monumental, but it didn't come without strife. Biases, prejudices, and underrepresentation remained commonplace. Rosalind Franklin's DNA structural work was credited to her male colleagues, underlining the gender bias. Her story stands as a stark reminder of the battle women had to, and continue to, wage in STEM fields despite the significant strides.

Yet, these stories of triumph amidst turbulence never failed to inspire. Against all odds, these women prevailed, their work reverberating across generations and breaking barriers of historically male dominance.

3.4. An Unceasing March

In spite of the challenges that remain, women in STEM have persisted, their narratives echoing resilience and defiance. From

Shirley Ann Jackson, the first African-American woman to earn a doctorate at MIT, to Ada Yonath, a crystallographer who astonished the science community with her research on the structure of the ribosome, the march towards equality has been unceasing.

The victories are indeed worth celebrating. Yet, the fight is far from over. The lens now shifts towards addressing the deep-seated gender bias in STEM education and employment to ensure more women can breach these barriers and secure their place in this rapidly evolving sector.

In conclusion, 'Breaking Barriers: The Triumph Over Historically Male Dominance' interprets the long journey involving persistence and determination by women in the STEM domain. The chapter is a powerhouse of inspirational stories, revealing the tale of brilliance and courage demonstrated by women in overcoming herculean hindrances.

Chapter 4. The Glass Ceiling: An Obstinate Reality In STEM

The bitter truth about our society is that despite strides towards gender equality, a treacherous glass ceiling looms above. This chapter explores the insidious nature of this barrier, its history, its manifestations, and its profound impact on women in the STEM sphere.

4.1. Unraveling the Concept: The Glass Ceiling

The term "glass ceiling" was first used in a 1984 Adweek article, referring to the unseen, unbreachable barrier that keeps women from rising to the ranks traditionally held by men, irrespective of their qualifications or achievements. Essentially, it is the invisible upper limit in corporations and other organizations, beyond which women, and often people of color, rarely advance.

What makes this oblique barrier so deceptively lethal is its lack of visibility, unlike formal barriers, such as exclusion from hiring for certain positions or explicit wage discrimination. While the latter forms have been somewhat addressed with legislation over the years, the glass ceiling persists, artfully camouflaged by notions of meritocracy, corporate culture, and societal expectations.

4.2. Historical Context

Historically, the entry of women in any professional realm has been marred by gendered constraints. Despite laws promoting equal opportunity and paycheck fairness, deeper layers of these constraints persist. This chapter, within its length constraints, delves into a vivid historical account, tracing the journey of systematic exclusion to a covert form of gender bias lurking in the recesses of STEM fields, fueling the glass ceiling phenomenon.

4.3. Gender Stereotypes: Fuelling the Fire

Deeply ingrained societal beliefs foster the unequal representation of women in STEM fields. The age-old stereotypical assumption that men are innately more capable in mathematics and science silently bolsters the argument against gender parity in STEM. Women are often subjected to the 'Stereotype Threat,' a psychological phenomenon where, due to societal expectations, women inadvertently underperform, thus feeding the stereotype cycle. Countering this corrosiveness demands an attitudinal shift at every social stratum.

4.4. Biases in the Workplace: Subtle yet Significant

The workplace is rife with covert biases. Promotion processes, for instance, often hinge on informal networks from which women are inadvertently barred because of conventional societal roles. Similarly, evaluation processes are mostly undefined, leaving room for unconscious biases to creep in and influence decisions concerning promotion and work assignments. The impact of these biases is not just frustratingly prevalent, but substantial enough to stifle progression and relegate talented women to perpetual mid-management positions.

4.5. The Vicious Circle: Few Women in Top Positions Propagate the Trend

The paucity of women at the top begets a self-perpetuating vicious circle. It not only maintains the dominance of the patriarchal structure but also leads to fewer women role models for the upcoming generation. This vicious circle not only aggravates the glass ceiling problem but makes rectification a daunting task.

4.6. The Economic Impact: Pay Gap and More

The gender pay gap is an enduring symbol of the economic manifestation of the glass ceiling. Despite exemplary expertise and skills, women in STEM often find themselves underpaid when compared to their male counterparts. This not only reflects the societal prejudices but also disincentives women from pursuing STEM careers.

4.7. Charting Forward: Potential Measures

The path to dismantling the glass ceiling is multifold. Promoting mentoring schemes, encouraging transparency in hiring and promotion processes, implementing objective assessment parameters, and fostering a more balanced work-life culture could serve as a good starting point. Policies to mitigate unconscious biases both at the workplace and in our societal structure are imperative.

In conclusion, the enigma of the glass ceiling is a deep-seated societal construct and thus, a daunting one to address. However, a concentrated effort from all shareholders could put an end to its existence, and herald an era of unprecedented growth, diversity, and equality in STEM fields. A world where promising talent isn't quashed by invisible ceilings, but bolstered by strong pillars of inclusion, support, and fairness.

Chapter 5. The STEM Gender Gap: Dissecting Its Origins and Effects

The issue of gender inequity in STEM professions commences from a profound and multilayered space. This long-standing gender gap can be held accountable to a wide array of factors ranging from societal norms, individual biases, to systemic institutional factors, resulting in a significant underrepresentation of women in STEM fields. Given the complexity of this issue, it becomes particularly important to dissect its origins and evaluate its effects, thereby paving the way towards sustainable solutions that intend to bridge this gap.

5.1. Historical Context of Gender Bias

Historically, scientific and technical fields have been persistently gendered, with societal norms associating masculine traits with the prerequisite aptitudes and personalities necessary for scientific and technical work. The persistence of such biased norms has resulted in the sidelining of women's capabilities and contributions in STEM fields. Women pioneers who made significant contributions to their respective fields often found their accomplishments marginalized, unrecognized, and even assigned to their male colleagues, leading to a notable skew in the gender distribution of innovation and advancement in STEM fields.

Objectively acknowledging and studying this historical context is not only essential for understanding the origination of the STEM gender gap, but it further serves as a reminder of the need to consciously combat such biases in the present times.

5.2. Socialization and Its Effect on Interest Formation

Gender socialization, which begins at early childhood, plays a pivotal role in shaping interest and proficiency in STEM subjects. Traditional norms and societal expectations often subtly direct girls away from math and science, and instead push them towards more 'feminine' fields.

Numerous studies pointed out that teachers, parents, and even educational toys prevalently made for girls unintentionally propagate stereotypes, thus curbing girls' interest in STEM. Furthermore, popular culture overwhelmingly showcases scientists and engineers as male, thereby further discouraging girls from envisioning a future for themselves in STEM.

5.3. Gender Stereotypes and Bias in Educational Institutions

Educational institutions often fail to rectify and may even perpetuate gender stereotypes that suggest inherent inferiority of women in STEM subjects. Academic curricula, unintentionally or otherwise, often echo such biases, creating discouraging environments for women pursuing STEM.

Many studies have documented the underrepresentation of women in advanced STEM courses and majors, suggesting an insidious bias against women. Such biases may lead to self-fulfilling prophecies, where women choose non-STEM fields due to their perception of unwelcome or hostile environments in STEM disciplines.

5.4. Impacts of the STEM Gender Gap

The underrepresentation of women in STEM fields has severe repercussions spanning socio-economic to innovation realms. From a macroeconomic perspective, the gender gap in highly lucrative STEM fields perpetuates the gender wage gap, contributing to wider societal inequities.

Moreover, the lack of diversity in STEM fields may hamper the quality and inclusivity of innovations. A homogenous group of innovators tend to solve problems from a limited perspective, potentially missing out on innovative approaches that a more diverse group might bring.

5.5. Conclusion: A Call to Bridge the STEM Gender Gap

Understanding the deep-seated origins and the far-reaching impacts of the STEM gender gap highlights the urgency of the need for targeted interventions. It becomes crucial to address the implicit and explicit biases prevalent at all levels, from societal to institutional, to ensure a more fair, inclusive, and innovative STEM landscape for future generations.

The path forward will require consistent efforts from educators, policy-makers, and society at large to deconstruct biases, encourage diversity and inclusivity, and to ensure every bright mind has the opportunity to shine in the world of STEM, irrespective of gender.

Chapter 6. Role Models and Mentors: Their Pervasive Influence in STEM Career Paths

The influence of role models and mentors in any career trajectory is indisputable, and perhaps, no field demonstrates this more tangibly than the Science, Technology, Engineering, and Mathematics (STEM) disciplines. For centuries, these were domains largely populated by men, cultivated in a soil of traditional norms and legacies that often sidelined women. However, today, as we navigate the arguably more inclusive landscapes of the 21st century, looking back at the paved paths of tenacious women, and their mentors guiding their voyage, unveils an inspiring saga of audacity, wisdom, and persistence.

6.1. The Role Model Effect: Inspiring Generations

Role models function as lighthouses, illuminating pathways in which potential seems not just possible but also achievable. Women in STEM have had their journeys marked by numerous such inspiring figures - pioneers who have refused to be circumscribed by societal dictates, thus shattering long-standing stereotypes about women's participation in these fields. Certainly, a narrative of their grit and determination has been key in influencing young sparks of talent to dare to dream and aspire.

Marie Curie, a symbol of stalwart resilience, awarded twice with the Nobel Prize, perfectly exemplifies the imprint that role models can leave for later generations. Rosalind Franklin, who undeterred by the heavy shroud of gender bias, contributed significantly to the discovery of the DNA structure. Katherine Johnson, a 'computer' at the National Aeronautics and Space Administration (NASA), whose mathematical calculations set the foundation for numerous space missions, including the monumental Apollo 11. It is these trailblazers who have inspired many women to envision a future where they too can partake in the scientific endeavor, thereby redefining what it means to be a woman in STEM.

6.2. The Transformational Influence of Mentors

Mentors, on the other hand, are instrumental in providing direction, nurturing talent, fostering confidence, and imparting wisdom based on their own experiences, insights, and tribulations. These individuals can be influential figures during one's years of study or even throughout one's career.

Research has always shown a positive correlation between mentoring and career success in STEM. Mentoring relationships often yield more potent results when the mentor shares the mentee's gender, particularly for women, as these relationships facilitate rapprochement with potential role models and lend understanding to female-specific challenges in this field.

Noteworthy instances proliferate, underlining the impact of this mentor-mentee dynamic. For instance, the story of Dr. Mae Jemison and her high school teacher, who nudged Jemison's curiosity about the cosmos, undeterred by her black and female identity. Or the invaluable mentorship of Chien-Shiung Wu, a pioneering experimental physicist, guided by her professors back in China, and later at the University of California, Berkeley. Similarly, Sally Ride, the first American woman to journey into space, who acknowledged the significant influence of her parents and Stanford University professors in sculpting her astounding career.

6.3. The Empowerment of Peer Networks

It is noteworthy to identify the empowerment derived from extensive networks of peers who have been through similar experiences. Women in STEM often report that peer networks are critical in providing support, fostering resilience, and priming them for potential challenges. SWE (Society of Women Engineers), AnitaB.org, and AWIS (Association for Women in Science) are commendable examples of organized efforts to establish such empowering networks.

6.4. The Imperative of Representation: 'You Can't Be What You Can't See'

Fundamentally, the power of seeing someone who resembles oneself in a specific role or profession stimulates a sense of attainability, dismantling the barriers of doubt. Stereotypes and gender biases can only be effectively neutralized when counter examples become prevalent and popular. The more we witness women excelling in STEM, the stronger we counteract the inherent gender stereotyping in these fields.

Despite these strides, our society still has a long way to go to ensure equitable representation, and it is here that consistent efforts in maintaining and nurturing role models and mentors become ever essential. It's a complex issue and there's no instant solution; it will require effort on multiple fronts, from policy changes to early education intervention, encouraging more women to become visible mentors and setting an example for the next generation.

The stories of women in STEM are stories not only of personal achievement but are etched in the broader narrative of progress and inclusion, providing hope and inspiration to countless young women. They spotlight the importance that role models and mentors can have in charting a new course of discourse in STEM, significantly impacting ambitions and aspirations. And while the terrain may still be tough, the promise of progress can be glimpsed in the twinkling stars of determination and resilience lighting up an ever-expanding universe of possibilities. The more women we have trailblazing this path to innovation, the more it becomes an impending reality that every spark of potential can indeed become the light that illuminates the way forward.

Chapter 7. Indian Women in STEM: A Success Story

India is a land of myriad contrasts and unending paradoxes. A country that flows with the rhythm of its rich cultural history, it cherishes its time-honored traditions and yet, stages an arena for evolution, adaptation, and modernization. The narratives of Indian women in Science, Technology, Engineering, and Mathematics (STEM) unfold within the weave of such fascinating contradictions. This chapter endeavors to chronicle the journeys of these unstoppable women who, against all odds, have carved a niche for themselves in STEM, thereby becoming game-changers and inspiration sources.

7.1. Industry Overview and Background

Despite being home to one third of the world's illiterate population, India has shown commendable growth in its pursuit towards educational development. The mushrooming of technologically advanced institutes and the rising competency in international academic rankings narrate the story of a nation marching towards edification. The inclusion of women in this journey, particularly in STEM, has led to an interesting evolution of gender dynamics.

Historically, Indian society housed deeply ingrained gender-based stereotypes, with widespread acceptance that men were innately better suited for scientific pursuits. However, the 21st century ushered in a new era where such stereotypes are being vehemently challenged, and we see an ever-growing number of women embarking on STEM career paths.

7.2. Pioneering Presence: Women Who Shattered The Stereotypes

As we trace the history of Indian women in STEM, we stumble upon numerous names that have broken the conventions and shattered the glass ceiling. Historically, the hallowed halls of STEM were predominantly occupied by men, but these women defied societal presumptions and became trailblazers in their respective fields.

One such woman was Janaki Ammal. Born in 1897 in the southern state of Kerala, Ammal was one of the first women scientists to gain recognition on an international platform. Renowned for her work in cytogenetics and phytogeography, she even has a flower named after her. Her relentless pursuit towards knowledge, despite living in a deeply patriarchal society, set the precedent for the women who would follow in her footsteps.

Rajeshwari Chatterjee was another such revolutionary woman who emerged from the realms of theoretical science and applied engineering. As the first woman engineer from Karnataka, she fought against societal norms that questioned the place of a woman in a technical world. Her illustrious career at the Indian Institute of Science and her efforts towards developing microwave engineering paved the way for the women who would dare to dream of working with machines and circuitry.

7.3. Present Scenario: Growth, Trends and Challenges

In the present scenario, the landscape of Indian women in STEM has significantly evolved and expanded. Women now constitute about 14% of the total 280,000 scientists, technologists, and engineers in research and development institutions in India, according to a 2020 report by UNESCO. This might seem like a small number in the grand scheme of things, but it nonetheless represents great leaps of progress in a traditionally male-dominated arena.

However, even with such progressive trends, the journey is not devoid of challenges. Women in STEM face multiple levels of adversity, from gender bias to the hurdles of combining a high-pressure career with familial responsibilities. Preference for male colleagues in project teams, maternity-related career breaks leading to a slower career progression, lack of mentorship and role models – these are just a few of the obstacles that make their journey in STEM an uphill climb.

7.4. Way Forward: Policies, Initiatives, and Future Prospects

India has undertaken considerable efforts to encourage and facilitate the participation of women in STEM fields. Policies and programs, such as the "KIRAN" (Knowledge Involvement in Research Advancement through Nurturing) initiative by the Department of Science and Technology (DST), aim to bring gender parity in the science and technology sector. The Indian government, NGOs, and even corporate entities are playing instrumental roles in improving the gender ratio in STEM fields through scholarships, mentorship programs, and providing platforms to showcase talent.

The march of Indian women in STEM is undeniably a success story, worthy of being celebrated and lauded. However, it should also serve as a reminder that there is still a long road ahead. The future rests in continuing these initiatives and constantly chipping away at the obstacles in order to create a holistic, inclusive environment that fuels growth, encourages diversity, and ultimately, leads to a richer, more robust landscape of scientific advancement and technological innovation.

In the end, it is not just about the success of Indian women in STEM, but about how their journey changes mindsets, redefines societal norms, and creates a legacy for future generations to come. For Indian women daring to venture into STEM, this is an ode to their resilience, determination, and courage - a testament to their tenacity and a must-read for every aspirant seeking inspiration.

Chapter 8. African Queens of STEM: From Cape To Cairo

The continent of Africa, rich in culture, resources, and potential, has been blazing its own unique trajectory on the globe's STEM map. From Cape Town at its southern tip to the bustling streets of Cairo in the north, African women are leaving an indelible mark in Science, Technology, Engineering, and Mathematics fields. They embark on STEM careers armed with knowledge, innovation, and more than anything else, a resilient spirit that keeps them marching forward.

8.1. The Illuminating Torchbearers

The role of African women in STEM is much like that of a torchbearer. They carry the light of knowledge and innovation, bringing forth a new era of African-led STEM advancements. Many have gallantly taken up the challenge, sometimes even recklessly breaking traditional conventions.

Dr. Amina Abubakar, a Kenyan pharmaceutical scientist, is one such trailblazer. Despite the prevailing belief of a woman's role as a homemaker, Dr. Abubakar pursued an education in pharmacy and therapeutics. Her groundbreaking work in antiviral therapeutics, specifically HIV treatments, draws praise and admiration from the global scientific community. She broke free from societal expectations and courted the audacious idea to better the world while anchoring her roots firmly in the African soil.

Similarly, Dr. Doyin Odubanjo, a Nigerian public health professional, has been instrumental in the battle against infectious diseases like malaria and ebola. Through aggressive research and advocacy, Dr. Odubanjo highlights the need for Africa's healthcare systems to be self-reliant and robust. Her contributions have transformed health policymaking, altering the face of Africa's public health paradigm.

These torchbearers, among many others, signify the pioneering spirit of African women in STEM fields. They have not only challenged the status quo but have also ignited a beacon of hope for aspiring STEM enthusiasts.

8.2. Confronting the Whirlwind: Challenges and Barriers

The path towards any significant achievement is seldom smooth. For women in Africa, pursuing a career in STEM fields often involves confronting pervasive socio-cultural biases and systemic barriers. One of the most pressing issues is education accessibility, or rather, its scantiness.

Girls in numerous African regions often face restrictive cultural norms and socioeconomic limitations that hinder their access to quality education, particularly in STEM subjects. Early marriages, biased attitudes towards girls' education, and the lack of female role models in STEM subjects only serve to exacerbate the situation.

Addressing such deeply rooted issues requires thoughtful and concerted efforts from institutions, governments, and society at large. These include creating inclusive education policies, enabling scholarships and funding for underprivileged girls, and raising awareness about the significance of girls' education in STEM fields.

8.3. Shaping the Future: Reformative Measures and Initiatives

Initiatives aimed at championing girl's education and promoting women's participation in STEM have been gaining traction in Africa. Organizations like African Women in Science and Engineering

(AWSE), Women in STEM Nigeria (WISN), and many others work relentlessly towards bridging the gender disparity. They offer mentorship programs, scholarships, and training sessions specifically targeting young African women ready to plunge into STEM careers.

Moreover, governments across the continent are also taking significant strides towards gender parity in STEM. For instance, South Africa's National Research Foundation offers a host of scholarships for female scholars in STEM. Meanwhile, Rwanda's commitment to achieving gender parity in education by 2020 is a testament to proactive governmental efforts.

8.4. Stories of Success: Blossoming Despite the Odds

Despite the hardships, African women in STEM have been registering notable victories over the years. Their achievements echo through the walls of innovation and inspire future generations.

Rania El Shatby, an Egypt-based marine biologist, stands tall at the helm of this success brigade. Not only was she named 'African Researcher of the Year' in 2019, but her decisive role in the discovery of a new Red Sea algae species has also put her at the forefront of marine conservation.

Further down in South Africa, Faith Osier, an immunologist of Kenyan origin, is known for her ground-breaking malaria research. As the first African woman president of the International Union of Immunological Societies, she represents the triumph of perseverance and the dawn of a new era of African women leadership in global scientific communities.

The stories of African women making inroads in STEM are intriguingly endless. As torchbearers, they continue to illumine the path for others, warming the ground for Africa's next STEM revolution. With every setback, they counter with an even stronger comeback. Challenges are met with resilience, and successes shine like lighthouses guiding the future prospects of African Queens of STEM. These narratives serve as reminders that beyond the barriers, lies a realm filled with endless possibilities. They are as much about the journey as the destination.

Chapter 9. Putting Latinas on the STEM Map: Triumphs and Challenges

Despite the underrepresentation of Latinas in Science, Technology, Engineering, and Mathematics (STEM) fields, numerous inspiring stories have emerged that epitomize perseverance and achievement. Embarking on this complex terrain marked with societal stereotypes, cultural barriers, and limited access to educational resources, these women have defied the odds to redefine the narrative, quietly yet impressively asserting their presence.

9.1. Building up the Numbers: The Current Landscape

While Latinas make up a significant portion of the U.S. population, they're conspicuously underrepresented in STEM fields, particularly in leadership profiles and patent holdings. According to the National Science Foundation, as of 2019, Latinas comprise only 2% of the STEM workforce, a stark contrast to their population proportion. Yet, these statistics don't tell the complete story.

Evidence suggests that Latinas are enrolling and graduating in STEM fields at an increasing rate, hinting at subtle but steady growth. Comprehension of these trends necessitates a closer look at a complex interplay of factors - a whirlwind of challenges, debates, triumphs, and systemic forces.

9.2. Challenging the Socio-Cultural Paradigm

Cultural norms and societal standards often morph into formidable barriers that inhibit Latinas from pursuing STEM careers. Common stereotypes, such as the erroneous belief that Latinas are uninterested or incapable of grasping complex technological and scientific concepts, contribute to a prevailing perception bias. These biases, internalized and propagated, generate a self-perpetuating cycle of discrimination and disadvantage.

Additionally, language barriers and the pressure to conform to traditional gender roles amplify these challenges. In many Latin cultures, women are expected to prioritise familial responsibilities over career ambition, which can limit their pursuit of STEM academic paths.

9.3. Unveiling the Triumphs

Amidst the struggles, numerous Latin American women have found success in STEM, serving as beacons of hope and perseverance. Notable figures such as Ellen Ochoa, the first Hispanic woman to go to space, and Lydia Villa-Komaroff, a molecular biologist known for her work in insulin synthesis, are stellar examples of Latinas asserting their prowess in STEM.

Moreover, an increase of Latinas in influential roles, such as Patricia Garcia, appointed as Peru's first female Minister of Health and researcher in public health, indicates a progressive shift.

9.4. Bridging the Gap: An uphill Task

The path to increasing Latina representation in STEM is challenging but increasingly seen as achievable. Significantly, the focus has broadened from simply encouraging Latinas to consider STEM to ensuring they possess the skills and resources required to thrive in these fields.

Educational scholarships, targeted mentorship programs, and more inclusive hiring practices are prevalent approaches. Further measures include addressing in-school biases, providing language support, and offering opportunities for internships and research.

9.5. Looking Forward: A Pledge for Inclusion

Latina representation in STEM is a multifaceted issue interwoven with race, gender, and socio-economic inequities. Yet, each achievement charts a course for progressive change, inspiring the next generation of Latinas to dream bigger.

There is a push towards inclusivity that needs to be strengthened, ensuring visible role models and mentors are highlighted, and pathways to employment are more accessible. All stakeholders—educators, parents, policymakers, corporations—must collaborate to tear down the long-standing walls of gender and racial discrimination that continue to hinder Latina progress in STEM.

Integrating Latinas in STEM is more than a targeted demographic initiative. It is a crucial step towards bringing diverse perspectives, fostering innovation, and driving societal progress. The future of STEM needs Latinas. Their talent, creative aptitude, and determination deserve recognition and amplification. Their triumphs paint a vibrant picture of what potential lies ahead, the myriad of opportunities and accomplishments that may unravel when diversity in STEM is not an exception but a norm.

In conclusion, 'Putting Latinas on the STEM Map: Triumphs and Challenges' delineates a vivid illustration of a complex equilibrium of attributable victories and the grand hurdles still yet to be conquered. It is a continuing journey, but with every step taken, the strides towards achieving true parity in STEM become more notable. With every story of success, a new beacon illuminates the path for countless nascent dreamers waiting to claim their rightful place in the realm of STEM.

Chapter 10. Future Endeavors: Emerging Fields And Opportunities For Women In STEM

The advancing realms of technology and scientific discovery are constantly brimming with fresh and fascinating opportunities for those keen enough to embrace them. For women in STEM, this horizon of emerging fields and opportunities presents a well-lit path brightly glowing with possibilities that can be tapped into.

10.1. The Advent of Artificial Intelligence and Machine Learning

With the profound technology revolution that our epoch endures, Artificial Intelligence (AI) and Machine learning (ML) are on the frontlines; shaping the future trajectory of numerous industries, from healthcare and finance to transport and entertainment. Characterized by unequivocal potentialities, these domains stand as fertile grounds for women in STEM to inscribe their own trajectories. Here, insights can be daringly explored, and conventional rules can be dynamically and efficiently rewritten.

Harnessing AI and ML's enigmatic capabilities, women are currently engineering sophisticated systems and algorithms that revolutionize predictive modeling, facilitate effective decision-making processes, and drive profitable business outcomes. In this highly advanced domain, every innovation, every leap, every stride taken has the potential to open up new frontiers and reshape pre-existing narratives fortifying the space for women's significant contributions.

10.2. The Rise of Data Science

The influx of big data has necessitated the evolution of data science. The distinct blend of statistical analysis, coding proficiencies, and investigative acumen makes this space ripe for STEM-wielded women to explore. The gender disparity in this sphere is stubborn, yes, but it also presents profound opportunities for rectification by creating job spaces that demand diverse perspectives and innovative strategies, precisely the kind that visionary minds of women can provide.

Women as data scientists can contribute significantly to business intelligence initiatives, such as customer relationship management, fraud detection, or enhancing operational efficiency. They can precisely interpret complex algorithms, generate predictive models, and craft data-driven business recommendations. The opportunity for tangible progress in this domain is vast, and the potential impact for women is equally substantial.

10.3. The Revolutionary Biotechnology and Genomics

Biotechnology and genomics are sprawling fields characterized by continuous advancements that promise unprecedented opportunities for women in STEM. Pioneering women are currently expanding the boundaries of molecular biology, gene therapy, and personalized medicine by combining their scientific acumen with emerging genomic technologies.

In this multifaceted arena, women are pushing the envelope, undertaking potentially transformative research, and creating breakthrough solutions that have pivotal implications for human health and the overall enhancement of life. This segment presents a unique intersection of biology, technology, and medicine, generating an array of opportunities for women to advance scientific and technological landscapes.

10.4. The Expanding Universe of Cybersecurity

As our world has increasingly digitized, guarding our digital assets has become an urgent need, thus belying the tremendous opportunity in the field of cybersecurity. The role of women here is undeniable. Women in cybersecurity bring diverse perspectives, novel approaches to problem-solving, and enhance team dynamics, all of which are critical aspects in responding to cyber threats effectively.

Challenging the traditionally male-dominated industry, women in cybersecurity are altering the narrative by punctuating the space with their commensurate skills and innovative competencies. The door to this burgeoning field is wide open for enterprising women, armed with their analytical prowess and technical skills, to cross thresholds and redefine paradigms.

10.5. The Ever-evolving Domain of Quantum Computing

Quantum computing, arguably one of the most exciting facets of modern science, holds immense promise for women in STEM. This realm, still in its nascent stage, offers boundless opportunities for innovation and exploration. Women are playing groundbreaking roles in harnessing and taming quantum systems, thus significantly influencing the trajectory of quantum science and engineering.

From resolving complex computations impossible for classical computers, to problem-solving in logistics and cybersecurity, the opportunities are as diverse as they are enticing. The exponential processing power of quantum computers beckons aspirational women to leverage this intriguing paradigm and etch their narrative in the annals of quantum lore.

In conclusion, as we stand on the cusp of a new era in science and technology, it is undeniable that numerous emerging fields are primed for exploration. This offers a broad arena for women in STEM to strive and thrive. The opportunities for women in these burgeoning sectors are manifold, and the potential for significant impact is undeniably high. The fortitude of the women forging paths in these innovative domains, serves as an inspiration to the future generation of female scientists, technologists, engineers, and mathematicians.

Chapter 11. Creating Inclusive Spaces: Recommendations for Encouraging Women in STEM

Creating inclusive spaces is a multifaceted and complex endeavor, one that calls for an interlocking cogs approach, with all players - teachers, institutions, corporations, communities, and governments - functioning in unison towards the desired outcome. The vision is to nurture and cultivate an environment where women feel encouraged and supported to pursue and thrive in STEM.

11.1. Encouraging Early Interest in STEM

The first stepping stone to creating a more inclusive ecosystem lies in fostering an early interest in STEM among young girls at the grassroot level. Scholars believe that incorporating STEM activities that resonate with their interests is an effective strategy. STEM toys such as Lego Mindstorms, chemistry sets, discovery kids' telescopes, etc., can stimulate curiosity and engagement, making learning an enjoyable and rewarding experience. In schools, educators can diversify teaching methods by incorporating hands-on practical sessions, lively discussions, and problem-solving exercises. This approach introduces and cultivates a sense of excitement about the spaces they generally find intimidating or discouraging.

11.2. Addressing Stereotypes and Biases

A significant roadblock deterring women's participation in STEM arises from established stereotypes and biases. Institutions must actively work towards disrupting such narratives using a two-pronged approach: combating unconscious stereotypes and confronting overt biases. This could involve unconscious bias training for faculty, sensitizing young minds to gender equality, and implementing strict zero-tolerance policies against discrimination and biases. Moreover, it is imperative to properly address even the subtle forms of bias, like 'microaggressions,' which, when overlooked, have the potential to create a hostile learning environment.

11.3. Role of Mentorship

Mentorship plays a crucial role in influencing career decisions and advancements, particularly in STEM fields. Creating mentorship programs wherein women at various stages of their STEM journey can interact, share their narratives, and learn from others' experiences is a potent tool. Senior women mentors can provide invaluable insights, advice, and perspective to their junior counterparts, helping them to navigate the challenging terrain of STEM.

11.4. Government and Scholarly Interventions

Governments worldwide have a critical role in shaping educational policies. Legislation should encourage initiatives that promote the involvement of women in STEM, such as grants to facilitate research, scholarships for female students, and awards recognizing women's contributions in STEM fields. Moreover, scholarly research should be emphasized to continually update the understanding of gender dynamics in STEM, thus equipping policymakers with detailed and scientifically-backed evidence for effective decision-making.

11.5. Inclusive Hiring Practices and Job Security

Companies can also contribute significantly to creating an inclusive STEM environment through their employment policies. Embracing diversity when hiring and creating flexible work conditions are some steps that can ease the transition for women entering the workplace, and help retain those who are already there. Another integral part is creating a safe, inclusive workspace free of any form of discrimination - an environment where women can freely express their ideas and perspectives.

11.6. Bridging the Confidence Gap

Research indicates the existence of a 'confidence gap' between men and women in the STEM fields, often leading women to doubt their abilities and underplay their achievements—a phenomenon known as the 'Impostor Syndrome.' To tackle this, workshops and training sessions aiming at building self-confidence and combating Impostor Syndrome should be organized. Also, lauding and publicising the successes of women in STEM can serve as powerful motivation.

In conclusion, 'Creating Inclusive Spaces' is a collective endeavor, one that requires concerted efforts from institutions and individuals alike. It is going to take time, resources, continued research, and most importantly, a widespread commitment to the cause to achieve gender parity in STEM. But the results - a richer, broader talent pool, fresh perspectives and ideas, and myriad innovative solutions to complex problems - undoubtedly make it more than worth the effort.

From the homes and classrooms to the highest echelons of power, everyone has a role to play and a contribution to make. Together, we can ensure a brighter, more inclusive future for women in STEM.

www.ingramcontent.com/pod-product-compliance
Lightning Source LLC
Chambersburg PA
CBHW070951220526
45471CB00007B/2982